The Book of
100
Equations
and other Algebra Struggles

The Algebra that Almost Kills Calculus

Introduction

Having worked with many many college level calculus students, I have witnessed repeatedly the common weaknesses among these students. Although it is true that exponentials, logarithms and trigonometry are among the common struggles, students still struggle in calculus after taking a traditional precalculus course that addresses exactly these topics. The biggest challenges are not the big topics previously listed but rather, algebraic manipulation. Manipulating variables, formalizing patterns, and solving equations top the list. Sadly, in most pre-calclulus books, attention to these troubles are contained to a single preliminary chapter at the beginning of the book and in most calclulus books as an appendix.

This workbook has been written to address the top struggles I have encountered as a professor of mathematics over the last ten years. It is divided by the following

I. The list of 100 equations solved in this book divided into their respective sections.

II. Common (oh so common) algebra errors

III. The techniques described and illustrated in each section type, the definitons needed for the equations in each section and the statement of each equation with space for the reader to attempt to solve each equation on his/her own.

IV. The detailed solutions to each equation.

To contact the author

Email her at elizabethstepp@gmail.com
or
Visit her site at www.elizabethstepp.com

Part I The List

Can you solve each of these?

A. Linear Equations and Absolute Value Equations

    1. $4(x - 1) - (3 - x) = 2$

    2. $1 - [3 - [4 - [5 - x]]] = 6(2 - x)$

    3. $-3p + 4 = -2(1 - p)$

    4. $3x - 2 - (x - 4) = 6(-(3 - 2x))$

    5. $|2x| = 4$

    6. $|3 - (x - 2)| = 6$

    7. $-2|3(x - 1)| = -4$

    8. $|3 - x| = |2x + 1|$

B. Equations with Fractions

    1. $\dfrac{1}{2}x + \dfrac{2}{3}x - \dfrac{3}{4}x = 2$

    2. $\dfrac{1}{2}(x - 2) + 3x = \dfrac{3}{2}(2 - x)$

    3. $\dfrac{2}{3}x - \dfrac{3}{4}(2 + x) = \dfrac{1}{6}(x - 1) + x$

    4. $\dfrac{1 + \dfrac{1}{x}}{x - \dfrac{x}{2}} = 3$

    5. $\dfrac{\dfrac{1}{2} - \dfrac{3}{4}x}{\dfrac{2}{3x} + \dfrac{5}{6}} = 0$

C. Solving for a Variable Amongst a Sea of Letters

    1. Solve for a; $2a^2 - 3b^2 = c^2$

    2. Solve for M; $F = G\dfrac{mM}{r^2}$

    3. Solve for h; $S = 2\pi r^2 + 2\pi rh$

    4. Solve for y; $x = \dfrac{1 - y}{y}$

    5. Solve for d; $2acd + 3(c - d) = a + 2(b - d)$

    6. Solve for $\dfrac{dy}{dx}$; $3x\dfrac{dy}{dx} - 2\left(1 - \dfrac{dy}{dx}\right) = 3 + x\dfrac{dy}{dx}$

    7. Solve for l; $\dfrac{1 + 3l}{l - 1} = \pi$

D. Basic factoring

    1. $x^8 - 1 = 0$

    2. $4x^2 - 36 = 0$

    3. $x^2 + 1 = 0$

    4. $(x + 2)^4 - (x + 2)^2 = 0$

    5. $x^6 - 9 = -1$

    6. Solve for all x and y that satisfy this equation: $6xy^2 - 9xy - 6x = 0$

E. Factoring Quadratic Equations

    1. $2x^2 + 3x - 2 = 0$

    2. $-3x^2 + x = -2$

    3. $x^2 - 7 = 0$

    4. $6x^2 - x = 2$

    5. $-2x^2 + 10x = 12$

F. Quadratic Equations by way of the Quadratic Formula

    1. $ax^2 + bx + c = 0$

    2. $3x^2 - 4x = \dfrac{1}{2}$

    3. $2x - 7 = -x - x^2$

    4. $px - qx^2 = sx^2 + 2$

    5. $\dfrac{1}{2}x^2 - \dfrac{2}{3}x - 2 = 0$

G. Completing the Square

    1. $x^2 + 4x - 7 = 0$

    2. $3x^2 - 6x = 9$

    3. $2x^2 - 3x = 5$

    4. $\dfrac{1}{3}x^2 - 2x + 1 = 2$

    5. Solve for x; $ax^2 + bx + c = 0$

H. Equations with Fractional or Negative Exponents

    1. $4(x - 1)^{1/2} - 3(x - 1)^{3/2} + (x - 1)^{5/2} = 0$

    2. $x^{1/2} + 3x^{-1/2} = 10x^{-3/2}$

    3. $(x - 1)^{1/2} + 3x(x - 1)^{-1/2} = 0$

    4. $(x + 2)^{4/5} - (2x + 4)^{-1/5} = 0$

    5. $3(x + 1)^{-1} - 3(x + 1)^{-2} = -2$

I. Radical Equations

1. $\sqrt{x-2} = 4$

2. $3\sqrt{x+1} = (3-x)\sqrt{x+1}$

3. $\sqrt{3-x} + 2 = \sqrt{x+1}$

4. $\sqrt{\sqrt{x+5}} - 5 = 0$

J. Rational Equations

1. $\dfrac{5}{x} + \dfrac{3}{x-2} = 4$

2. $\dfrac{1}{x-1} + \dfrac{x+2}{(x-1)^2} = \dfrac{1}{x}$

3. $\dfrac{3x^2 - 5x - 2}{x^2 - 4} = \dfrac{2}{3}$

4. $\dfrac{x+1}{(x-2)(3-x)} \cdot \dfrac{x-2}{(x+1)(2+x)} = 7$

5. $\dfrac{x}{3x-1} \div \dfrac{x^3}{(3x-1)(3x-2)} = 1$

K. Exponential Equations

1. $3^{2x-1} = 27$

2. $\left(\dfrac{1}{4}\right)^{-2} = 16^{-x}$

3. $2^x = 3^{x+1}$

4. $e^x + 6 = 3e^x$

5. $4e^{1-x} = 7$

6. $2^x 4^{-x} = 16$

L. Logarithmic Equations

1. $\log_2(3-x) = 3$

2. $2 + \log(x-2) = 4$

3. $\log_2(x^2 + 3x - 1) = 3$

4. $\log_3(x+1) - \log_3(2x+1) = 2$

5. $\log_2(\log_3(x-2)) = 2$

6. $\ln(1-x) + \ln(2x+1) = 0$

7. $3^{3\log_3 x} = 7$

8. $\ln(e^2 x + 3e + e) = \ln(2x) + 1$

9. $\log_3(x+1) + \log_2 8 = \log_3(3-x) + \log_5 25$

M. Trigonometric Equations

    1. $\cos x \sin x = \sin x$

    2. $4 \sin^2 x - 1 = 0$

    3. $(\tan x - \sqrt{3})(\sec x + 2) = 0$

    4. $\csc^2 x - 2 = 0$

    5. $\cot x - \sqrt{3} = 0$

    6. $2 \cos^2 x = \sin x + 1$

    7. $3 \sin 2x - 2 \sin x = 0$

    8. $\tan^2 x - 2 \sec x = 2$

    9. $\sec x \tan x - \cos x \cot x = \sin x$

    10. $\sin^2 x + 3 = 7 \cos^2 x$

N. Things that Look A LOT Like Quadratics

    1. $(x + 1)^2 - 5(x + 1) - 6 = 0$

    2. $2e^{2x} + 3e^x - 2 = 0$

    3. $\sin^2 x + \sin x = 2$

    4. $\sin^2 x = 2 \sin x + 3$

    5. $x^{1/3} + x^{1/6} - 2 = 0$

    6. $x^4 + x^2 = 2$

Keep Going...

Other Algebra Struggles

O. Inequalities

1. $\dfrac{3}{x} < \dfrac{1}{2}$

2. $2 \le 2x - 1 < 3$

3. $x^2 - 2x > 0$

4. $(x + 2)(x - 3) < -4$

5. $\dfrac{2 - x}{x + 1} > 4$

6. $\dfrac{(2 - x)(x + 1)}{3 - x} < 0$

7. $|2x - 1| < 3$

8. $|3 - x| > 2$

9. $x^2 - 3x - 10 \ge 0$

10. $2x^2 + 3x - 3 < -1$

P. Finding Equations of Lines (this will happen every day in calculus) and a bonus.

1. Find the equation of the line passing through the point $(2, 3)$ with slope $m = -2$.
2. Find the equation of the line passing through the points $(4, 5)$ and $(-2, 7)$.
3. Find the equation of the line passing through $(-2, 4)$ and parallel to the line $2x - 3y = 7$.
4. Find the equation of the line passing through $(3, -1)$ and perpendicular to the line $3x + 7y = -1$.

Bonus

Find the equation of the circle with center $(-2, 3)$ and radius $r = 2$.

1. $(a+b)^2 = a^2 + b^2$ ← Is so very bad but so very common.

   We know that $(a+b)^2 = (a+b)(a+b) = a^2 + ab + ab + b^2 = a^2 + \underline{2ab} + b^2 \neq a^2 + b^2$.

   The moral is: **Thou Shall Not Distribute Exponents Over Sums.**

2. $\sqrt{x^2 + y^2} = x + y$ ← Would be awesome if it were true but it isn't!

   Recall that $\sqrt{x^2 + y^2} = (x^2 + y^2)^{1/2}$ and now we see that this sin is really the first. You cannot simplify this one as is.

   The moral is: **Thou Shall Not Distribute Exponents Over Sums Even If the Exponents are Fractions Disguised as Square Roots.**

3. $\dfrac{x+2y}{2z} = \dfrac{x+\cancel{2}y}{\cancel{2}z} = \dfrac{x+y}{z}$ ← Is not as common - but common enough to mention.

   Note that $\dfrac{2x+2y}{2z} = \dfrac{\cancel{2}(x+y)}{\cancel{2}z} = \dfrac{x+y}{z}$ and you would never say

   $\dfrac{x+2y}{2z} = \dfrac{2x+2y}{2z}$

   The moral is: **Thou Shall Not Cancel Factors into Part of a Sum**

4. $\sin\theta\cos\theta = \cos\theta$ becomes $\dfrac{\sin\theta\cancel{\cos\theta}}{\cancel{\cos\theta}} = \dfrac{\cancel{\cos\theta}}{\cancel{\cos\theta}}$ becomes $\sin\theta = 1$. Why isn't this right??

   You can only divide both sides when $\cos\theta \neq 0$. But what about when $\cos\theta = 0$? By dividing by $\cos\theta$, you lose a potential solution. Instead, move everything to one side, factor, and use the Zero Product Property.

   $\sin\theta\cos\theta - \cos\theta = 0$

   $\cos\theta(\sin\theta - 1) = 0$

   $\cos\theta = 0$ or $\sin\theta - 1 = 0$. Now solve BOTH equations.

   The moral is: **Thou Shall Not Divide by Zero and Make Sure You Know What Zero May Look Like**

Part III - The Techniques, Examples and Practice

A. Linear Equations and Absolute Value Equations

Some Notes:

Solving linear equations is pretty straightforward at this level and actually the easiest of the types we solve in this workbook. The traps students fall into with these problems are generally contained to sign errors or distribution errors.

A couple of things to keep in your line of sight:

1. If we have a number multiplied to an entire sum, then remember to distribute the number to ALL terms inside the parenthesis. For example,

$$2(x + y - z) = 2x + 2y - 2z$$

2. This is also true with negatives. Remember that a negative preceeding a term, really means $-1\times$ that term.

$$-(x + y - z) = -x - y + z$$

The absolute value often complicates things. Recall the definition,

$$|x| = \begin{cases} x & \text{if } x \geq 0; \\ -x & \text{if } x < 0. \end{cases}$$

So when solving an absolute value equation, remember to consider both cases. For examples,

(a) $|x| = 7$ yields

$x = 7$ OR $x = -7$

(b) $|2x + 1| = 7$ yields

$2x + 1 = 7 \rightarrow 2x = 6 \rightarrow x = 3$ OR

$2x + 1 = -7 \rightarrow 2x = -8 \rightarrow x = -4$

(c) $3|x + 2| - 1 = 8$ First isolate the absolute value.

$3|x + 2| = 9$

$|x + 2| = 3$ yields

$x + 2 = 3 \rightarrow x = 1$ OR

$x + 2 = -3 \rightarrow x = -5$

Practice these ideas on the following few pages.

Practice Linear Equations and Absolute Value Equations

1. $4(x-1) - (3-x) = 2$

2. $1 - [3 - [4 - [5 - x]]] = 6(2 - x)$

3. $-3p + 4 = -2(1 - p)$

4. $3x - 2 - (x - 4) = 6(-(3 - 2x))$

5. $|2x| = 4$

6. $|3 - (x - 2)| = 6$

7. $-2|3(x-1)| = -4$

8. $|3 - x| = |2x + 1|$

B. Equations with Fractions

Some Notes:

It is often a good idea when encountering an equation that has fractions, to get rid of them! I don't mean just erase them, but rather, get rid of them by multiplying the entire equation by a common denominator, often the least common denominator (LCD). For example

$$\frac{1}{3}x - \frac{1}{2}(x-1) = \frac{5}{6}$$

The LCD is 6,

$$6\left(\frac{1}{3}x - \frac{1}{2}(x-1) = \frac{5}{6}\right)$$

Now cancel each denominator.

$$2(1)x - 3(x-1) = 5$$

Now solve the equation.

$$2x - 3x + 3 = 5$$

$$-x = 2$$

$$x = -2$$

Always go back and check as careless errors are often caught at this step.

Let's consider the case where we have fractions inside of fractions - oh my!

For example

$$\frac{\frac{1}{2}x - \frac{1}{3}}{\frac{1}{2} - \frac{3}{4}x} = \frac{1}{5} + \frac{1}{3}$$

First combine the numerator into one fraction, the denominator into one fraction and the RHS into one fraction.

$$\frac{\frac{3}{6}x - \frac{2}{6}}{\frac{2}{4} - \frac{3}{4}x} = \frac{3}{15} + \frac{5}{15}$$

13

$$\frac{\dfrac{3x-2}{6}}{\dfrac{2-3x}{4}} = \frac{8}{15} \quad \text{Cross mutliply}$$

$$\frac{3x-2}{6} = \frac{8}{15}\left(\frac{2-3x}{4}\right) = \frac{2(2-3x)}{15} = \frac{4-6x}{15}$$

$$\frac{3x-2}{6} = \frac{4-6x}{15} \quad (\text{LCD} = 30)$$

$$30\left(\frac{3x-2}{6} = \frac{4-6x}{15}\right)$$

$$5(3x-2) = 2(4-6x)$$

$$15x - 10 = 8 - 12x$$

$$27x = 18$$

$$x = \frac{18}{27} = \frac{2}{3}$$

Practice Equations with Fractions

1. $\dfrac{1}{2}x + \dfrac{2}{3}x - \dfrac{3}{4}x = 2$

2. $\dfrac{1}{2}(x-2) + 3x = \dfrac{3}{2}(2-x)$

3. $\dfrac{2}{3}x - \dfrac{3}{4}(2+x) = \dfrac{1}{6}(x-1) + x$

4. $\dfrac{1 + \dfrac{1}{x}}{x - \dfrac{x}{2}} = 3$

5. $\dfrac{\dfrac{1}{2} - \dfrac{3}{4}x}{\dfrac{2}{3x} + \dfrac{5}{6}} = 0$

C. Solving for a Variable Amongst a Sea of Letters

Some Notes:

Students get confused when letters are used to represent both constants and variables. Try and think of all constants as numbers and proceed the same way. For example, it would be easy to solve for x in the following

$$8 = \frac{5x}{3}$$

Multiply both sides by 3

$$8 * 3 = 5x$$

Divide both sides by 5

$$\frac{8 * 3}{5} = x$$

Apply these same principles to the following

Solve for t

$$P = \frac{Vt}{r}$$

$$Pr = Vt$$

$$\frac{Pr}{V} = t$$

Another example

Solve for t

$$A = Pe^{rt}$$

$$\frac{A}{P} = e^{rt}$$

$$\ln \frac{A}{P} = \ln e^{rt} = rt$$

$$\frac{\ln \frac{A}{P}}{r} = t$$

Practice Solving for a Variable Amongst a Sea of Letters

1. Solve for a; $2a^2 - 3b^2 = c^2$

2. Solve for M; $F = G\dfrac{mM}{r^2}$

3. Solve for h; $S = 2\pi r^2 + 2\pi rh$

4. Solve for y; $x = \dfrac{1-y}{y}$

5. Solve for d; $2acd + 3(c-d) = a + 2(b-d)$

6. Solve for $\dfrac{dy}{dx}$; $3x\dfrac{dy}{dx} - 2\left(1 - \dfrac{dy}{dx}\right) = 3 + x\dfrac{dy}{dx}$

7. Solve for l; $\dfrac{1+3l}{l-1} = \pi$

D. Basic Factoring

Some Notes:

Factoring and using the zero product property will be a technique a calculus student will use often.

The Zero Product Property (ZPP) - Given two real numbers a and b

$$a \cdot b = 0 \text{ then } a = 0 \text{ or } b = 0$$

Note there is **NO** not-zero product property. That is, if $a \cdot b = 2$, one can NOT conclude that $a = 2$ or $b = 2$. This only works with 0. Since you will often want to use the ZPP, make sure you first get your equation set equal to zero. Your goal then will be to factor your equation so that you can set each factor to zero and solve. For example,

$$x^2 + 3x - 28 = 0 \text{ factors to } (x + 7)(x - 4) = 0$$

$$x + 7 = 0 \rightarrow x = -7 \quad \text{or} \quad x - 4 = 0 \rightarrow x = 4$$

Consider this example

$$(x + 2)(x - 4) = -5$$

It is tempting to set each factor equal to -5, but don't! Remember there is no -5 product property. You must first multiply the left, add 5 and then factor again.

$$x^2 - 2x - 8 = -5$$

$$x^2 - 2x - 3 = 0$$

$$(x - 3)(x + 1) = 0$$

Now you can use the ZPP!

$$x - 3 = 0 \rightarrow x = 3 \quad \text{or} \quad x + 1 = 0 \rightarrow x = -1$$

Factoring Best Practices

(a) Check to see if there is a common factor for each term. If so, factor out the GCF and then procede to factor the rest if there is more factoring to do.

(b) Remember the difference of two squares: $a^2 - b^2 = (a + b)(a - b)$

(c) Remember that the sum of two squares DOES NOT factor.

(d) Remember the difference of two cubes: $a^3 - b^3 = (a - b)(a^2 + ab + b^2)$

(e) Remember the sum of two cubes: $a^3 + b^3 = (a + b)(a^2 - ab + b^2)$

Most Important: Factoring has a built in "check your work" feature. Always always re-multiply your factoring to check to see if you end up where you started. Many sign errors are caught this way. For a calculus student, there is no excuse for incorrect factoring.

Practice Basic factoring

1. $x^8 - 1 = 0$

2. $4x^2 - 36 = 0$

3. $x^2 + 1 = 0$

4. $(x+2)^4 - (x+2)^2 = 0$

5. $x^6 - 9 = -1$

6. Solve for all x and y that satisfy this equation: $6xy^2 - 9xy - 6x = 0$

Practice Factoring Quadratic Equations

1. $2x^2 + 3x - 2 = 0$

2. $-3x^2 + x = -2$

3. $x^2 - 7 = 0$

4. $6x^2 - x = 2$

5. $-2x^2 + 10x = 12$

E. Quadratic Equations by way of the Quadratic Formula

Some Notes: Not all quadratics factor! But if there is a solution to a quadratic, it can be found using the quadratic formula.

Let $ax^2 + bx + c = 0$ be a quadratic equation in standard form. That is, written in descending order and set equal to 0. Then the solution to this equation is

$$x = \frac{-b \pm \sqrt{b^2 - 4ac}}{2a}$$

Note that a quadratic equation has two distinct real solutions, one repeated real solution, or no real solutions. The discriminant: $(b^2 - 4ac)$ will tell us which.

(a) If $b^2 - 4ac > 0$ then the quadratic has two distinct real solutions.

(b) If $b^2 - 4ac = 0$ then the quadratic has one repeated solution.

(c) If $b^2 - 4ac < 0$ then the quadratic has no real solutions.

Solve $3x^2 - 2x = 4$

First get the quadratic in standard form

$3x^2 - 2x - 4 = 0$

Here $a = 3, b = -2$ and $c = -4$

$$x = \frac{2 \pm \sqrt{(-2)^2 - 4(3)(-4)}}{2(3)} = \frac{2 \pm \sqrt{4 + 48}}{6} = \frac{2 \pm \sqrt{52}}{6}$$

Now reduce: $x = \dfrac{2 \pm \sqrt{4 \cdot 13}}{6} = \dfrac{2 \pm 2\sqrt{13}}{6} = \dfrac{1 \pm \sqrt{13}}{3}$

Now go practice...

Practice Quadratic Equations by way of the Quadratic Formula

1. $ax^2 + bx + c = 0$

2. $3x^2 - 4x = \dfrac{1}{2}$

3. $2x + 7 = -x - x^2$

4. $px - qx^2 = sx^2 + 2$

5. $\dfrac{1}{2}x^2 - \dfrac{2}{3}x + 2 = 0$

F. Completing the Square

Some Notes:

Completing the square can come up when solving quadratic equations but it will also come up in calculus, so pay attention.

Here are the steps to complete the square in a quadratic equation:

1. Make sure the coefficient of the square term is 1. If it is not 1, then divide the entire equation by the coefficient.
2. Move the constant to the right side leaving on the x^2 term and the x term on the left hand side.
3. Now take the new coefficient of the x term and divide it by 2, square it and then add it to both sides of the equation.
4. The left side will not factor to a perfect square.
5. Take the square root of both sides and solve for x

What?? This is definitely best understood with an example.

$$x^2 + 6x - 2 = 0$$

1. The coefficient is 1.
2. $x^2 + 6x = 2$
3. $x^2 + 6x + 3^2 = 2 + 3^2 = 11$
4. $(x+3)^2 = 11$
5. $x + 2 = \pm\sqrt{11} \rightarrow x = -2 \pm \sqrt{11}$

Another example!

$$2x^2 + 3x + 2 = 0$$

1. $\dfrac{2x^2 + 3x + 2 = 0}{2} = x^2 + \dfrac{3}{2}x + 1 = 0$

2. $x^2 + \dfrac{3}{2}x = -1$

3. $x^2 + \dfrac{3}{2}x + \left(\dfrac{3}{2\cdot 2}\right)^2 = -1 + \left(\dfrac{3}{2\cdot 2}\right)^2$

   $x^2 + \dfrac{3}{2}x + \dfrac{9}{16} = -1 + \dfrac{9}{16} = \dfrac{7}{16}$

4. $\left(x + \dfrac{3}{4}\right)^2 = \dfrac{7}{16}$

5. $x + \dfrac{3}{4} = \pm\sqrt{\dfrac{7}{16}} = \pm\dfrac{\sqrt{7}}{4}$

   $x = -\dfrac{3}{4} \pm \dfrac{\sqrt{7}}{4}$

Practice Completing the Square

1. $x^2 + 4x - 7 = 0$

2. $3x^2 - 6x = 9$

3. $2x^2 - 3x - 5$

4. $\dfrac{1}{3}x^2 - 2x + 1 = 2$

5. Solve for x; $ax^2 + bx + c = 0$

G. Equations with Fractional or Negative Exponents

Some Notes:

Recall the rules for exponents. Let a, b and n be any real numbers.

$$a^m a^n = a^{m+n} \qquad \frac{a^m}{a^n} = a^{m-n} \qquad (a^m)^n = a^{mn}$$

$$(ab)^n = a^n b^n \qquad a^{-n} = \frac{1}{a^n} \qquad a^0 = 1, a \neq 0$$

These rules apply to *all* real numbers, including fractions. So,

$$a^{1/2} a^{2/3} = a^{1/2+2/3} = a^{3/6+4/6} = a^{7/6}$$

and

$$(a^{1/2})^{2/3} = a^{1/2 \cdot 2/3} = a^{4/6} = a^{2/3}$$

This is important to remember when factoring out fractional exponents. Perhaps we have $a^{4/5}$ and we wish to factor out $a^{-1/5}$. We must apply the multiplication rule in reverse.

$$a^{4/5} = a^{-1/5}\left(\frac{a^{4/5}}{a^{-1/5}}\right) = a^{-1/5}(a^{4/5--1/5}) = a^{-1/5}(a^{4/5+1/5} = a^{-1/5}a^{5/5} = a^{-1/5}a$$

Let's see this in action. But first lets look at a simpler example to guide us.

Factor out the GCF for $x^3 + 2x^2 - 3x = x(x^2 + 2x - 3)$.

Since x^1 is the greatest factor of x in common to all terms, this is our GCF.

Now condsier a harder similar example

$$x^{4/5} + 2x^{3/5} - x^{1/5}.$$

Here the greatest factor of x in common to all terms is $x^{1/5}$ so this is our GCF and factored we get

$$x^{1/5}(x^{4/5-1/5} + 2x^{3/5-1/5} - x^{1/5-1/5}) = x^{1/5}(x^{3/5} + 2x^{2/5} - x^0) = x^{1/5}(x^{3/5} + 2x^{2/5} - 1)$$

Notice that the greatest common factor in the above examples was the lowest power of x. Why?

So consider this example

$$3x^{3/2} - 6x^{1/2} - 9x^{-1/2}$$

The lowest power of x which is $x^{-1/2}$ and we also see that 3 is a common factor to each so our GCF is $3x^{-1/2}$.

$$3x^{-1/2}(x^{3/2--1/2} - 2x^{1/2--1/2} - 3x^{-1/2--1/2}) = x^{-1/2}(x^4/2 - 2x^{2/2} - 3x^0) = x^{-1/2}(x^2 - 2x - 3)$$

Pratice Equations with Fractional or Negative Exponents

1. $4(x-1)^{1/2} - 3(x-1)^{3/2} + (x-1)^{5/2} = 0$

2. $x^{1/2} + 3x^{-1/2} = 10x^{-3/2}$

3. $(x-1)^{1/2} + 3x(x-1)^{-1/2} = 0$

4. $(x+2)^{4/5} - (2x+4)^{-1/5} = 0$

5. $3(x+1)^2 - 3(x+1)^{-2} = -2$

H. Radical Equations

Some Notes:

Note that a radical is really an exponential power. For example, $\sqrt{x} = x^{1/2}$.

The formal definition given n, a non-zero integer.

$$\sqrt[n]{x} = x^{1/n}$$

This is a useful convention when solving equations that contain radicals. Solve for x.

$$\sqrt{x-1} = 8$$
$$\sqrt{x-1} = (x-1)^{1/2} = 8$$

To get rid of the exponent, raise both sides to the reciprocal power.

$$((x-1)^{1/2})^{2/1} = 8^{2/1}$$

$$x - 1 = 64$$

$$x = 65.$$

It is important to always check your solutions when raising equations to even powers. Sometimes extraneous solutions sneak in and we have to make sure not to let them!

$$\sqrt{65-1} = \sqrt{64} = 8. \text{ All is good with our solution.}$$

When solving an equation with a radical, it is a good idea if you can first isolate the radical first.

$$2\sqrt{2x+1} + 4 = 2x + 2$$

$$2\sqrt{2x+1} = 2x - 2$$

$$\sqrt{2x+1} = x - 1.$$

Now raise both sides to the 2 power.

$$2x + 1 = (x-1)^2 = x^2 - 2x + 1$$

$$x^2 - 4x = 0$$

$$x(x-4) = 0$$

$$x = 0 \quad \text{OR} \quad x = 4$$

Let's now consider an example that has two radicals.

$$\sqrt{x+1} - \sqrt{3-x} = 2$$

It is not possible to isolate both radicals so we will have to isolate each in separate steps. It is a common error to do this:

$(\sqrt{x+1} - \sqrt{3-x})^2 = (x+1) - (3-x) = 2^2$. This is incorrect! (See sin number 1).

Instead, isolate one radical and proceed cautiously. Follow me...

$$\sqrt{x+1} = 2 + \sqrt{3-x}$$

Now square each side - be careful.

$$(\sqrt{x+1})^2 = (2 + \sqrt{3-x})^2 = 4 + 4\sqrt{3-x} + 3 - x$$

$$x + 1 = 7 - x + 4\sqrt{3-x}$$

Now, isolate the second radical.

$$4\sqrt{3-x} = 2x - 6$$

$$\sqrt{3-x} = \frac{2}{4}x - \frac{6}{4} = \frac{1}{2}x - \frac{3}{2}.$$

$$3 - x = \left(\frac{1}{2}x - \frac{3}{2}\right)^2 = \frac{1}{4}x^2 - \frac{3}{2}x + \frac{9}{4}$$

Get rid of those fractions!

$$4\left(3 - x = \frac{1}{4}x^2 - \frac{3}{2}x + \frac{9}{4}\right)$$

$$12 - 4x = x^2 - 6x + 9$$

$$x^2 - 2x - 3 = 0$$

$$(x-3)(x+1) = 0$$

$$x = 3 \quad \text{OR} \quad x = -1.$$

Remember to check your answers to try and catch sneaky extraneous solutions.

When $x = 3$

$\sqrt{3+1} = 2 = 2 + \sqrt{3-3}$. We see, $x = 3$ is a solution.

When $x = -1$

$\sqrt{-1+1} = 0 \neq 2 + \sqrt{3+1} = 2 + \sqrt{4} = 4$. Thus, $x = -1$ is not a solution.

Note: $\sqrt{x}$ is the principle square root of x and is always non-negative.

$\sqrt{9} = 3$ whereas if $x^2 = 9$, then $x = \pm 3$. Different.

Practice Radical Equations

1. $\sqrt{x-2} = 4$

2. $3\sqrt{x+1} = (3-x)\sqrt{x+1}$

3. $\sqrt{3-x} + 2 = \sqrt{x+1}$

4. $\sqrt{\sqrt{x+5}} - 5 = 0$

I. Rational Equations

Some Notes:

Tackling rational equations is very much like working with equations that have constant fractions. Recall from the notes in the section on Equations with Fractions, when we have an equation with fractions, get rid of 'em! We did this by multiplying the entire equation by the LCD.

Let's start with an example

$$\frac{3}{x} + \frac{x-2}{x+1} = 5$$

Before we multiply by the LCD, it is important to first note the values x that cause our denominators to be zero. These values cannot be solutions to our equations.

In this example, we see that $x \neq 0$ and $x \neq -1$.

We are ready to multiply by our LCD $= x(x+1)$

$$x(x+1)\left(\frac{3}{x} + \frac{x-2}{x+1} = 5\right)$$

$$\frac{\cancel{x}(x+1)}{1}\frac{3}{\cancel{x}} + \frac{x\cancel{(x+1)}}{1}\frac{x-2}{\cancel{x+1}} = x(x+1)5$$

$$(x+1)3 + x(x-2) = 5x^2 + 5x$$

Now we use our previous techniques to finish.

$$3x + 3 + x^2 - 2x = 5x^2 + 5x$$

$$4x^2 + 4x - 3 = 0$$

$$x = \frac{-4 \pm \sqrt{16 - 4(4)(-3)}}{2(4)}$$

$$x = \frac{-4 \pm \sqrt{64}}{8} = \frac{-4 \pm 8}{8}$$

$$x = \frac{-4+8}{8} = \frac{4}{8} = \frac{1}{2} \quad \text{OR}$$

$$x = \frac{-4-8}{8} = \frac{-12}{8} = -\frac{3}{2}$$

Since both of these solutions are in the domain of our equation, they are our solutions! Of course it always a good idea to check your solutions as careless errors are often caught this way.

Now consider multiplying rational expressions.

Here are some good practices when facing the product of rational expressions.

(a) Factor all polynomials

(b) Find the domain of the equation

(c) Cancel and simplify common factors

(d) Multiply numerators and multiply denominators

(e) Solve the equation as discussed previously.

Let's see an example

$$\frac{x^2 - x - 2}{x^2 - 3x} \cdot \frac{x^2 - x}{x^2 - 1}$$

$$\frac{(x+1)(x-2)}{x(x-3)} \cdot \frac{x(x-1)}{(x+1)(x-1)} = 3.$$

We see that $x \neq 0$, $x \neq 3$, $x \neq -1$, and $x \neq 1$

$$\frac{\cancel{(x+1)}(x-2)}{\cancel{x}(x-3)} \cdot \frac{\cancel{x}\cancel{(x-1)}}{\cancel{(x+1)}\cancel{(x-1)}} = 3$$

$$\frac{x-2}{x-3} = 3$$

$$x - 2 = 3(x-3)$$

$$x - 2 = 3x - 9$$

$$7 = 2x$$

$$x = \frac{7}{2}$$

Dividing rational expressions is the same thing as multiplying them with one extra step. Recall that to divide by a fraction is to multiply by its reciprocal.

To divide rational expressions, we slightly ammend the list above

(a) **Change the division to multiplication by the reciprocal**

(b) Factor all polynomials

(c) Find the domain of the equation

(d) Cancel and Simplify common factors

(e) Multiply numerators and multiply denominators

(f) Solve the equation as discussed previously.

Practice Rational Equations

1. $\dfrac{5}{x} + \dfrac{3}{x-2} = 4$

2. $\dfrac{x}{x-1} + \dfrac{x+2}{(x-1)^2} = \dfrac{1}{x}$

3. $\dfrac{3x^2 - 5x - 2}{x^2 - 4} = \dfrac{2}{3}$

4. $\dfrac{x+1}{(x-2)(3-x)} \cdot \dfrac{x-2}{(x+1)(2+x)} = 7$

5. $\dfrac{x}{3x-1} \div \dfrac{x^3}{(3x-1)(3x-2)} = 1$

J. Exponential Equations

Some Notes: There are two main ways we will encounter exponential equations in calculus. One will involving get like bases and the other will involve the logarithm. Let's look at this in turn.

In the first case, if we can manipulate our equation to get it into the form $a^x = a^y$ then we can conclude that $x = y$. The key to note is that when our bases are the same, the exponents must be the same. Consider the following examples.

$$3^x = 9^{3-x} = (3^2)^{3-x} = 3^{6-2x}$$

Now that we have like bases, we can set their exponents equal

$$x = 6 - 2x$$

$$3x = 6 \rightarrow x = 2$$

Another example

$$\left(\frac{1}{3}\right)^2 = 27^x$$

It will be important to remember your rules for exponents as we summarized in a previous section.

Since $\frac{1}{3} = 3^{-1}$, our equation becomes

$$(3^{-1})^2 = 3^{3x}$$

$$-2 = 3x \rightarrow x = -\frac{3}{2}$$

Now we example exponential equations that require the use of logarithms to solve. Recall that

$$a^x = y \iff \log_a y = x$$

You should get comfortable going back and forth between the two forms.

$$2^x = 4 \iff \log_2 4 = x$$

Notice here we solved for the exponent by switching to logarithmic form.

$$\log_2 8 = x \iff 2^x = 8 \rightarrow x = 3$$

Let's use these skills to solve an exponential equation.

$$3^x + 7 = 11$$

First isolate your exponential

$$3^x = 4$$

Here we see it is difficult to get like bases so we use the logarithm

Switching to logarithmic form

$$\log_3 4 = x \quad \text{Bam.}$$

One of our favorite exponetial functions in calculus is $y = e^x$. To convert this to logarithmic form, we get, $\log_e y = x$. It is common practice in calculus to notate the "log base e of x" as $\ln x$. So rewriting the expression,
$$e^x y \iff \ln y = x$$

We need to recall our Properties for Logarithms

Given $a > 0$ and $M, N > 0$,

1. $\log_a M + \log_a N = \log_a MN \leftarrow$ The Product Property of Logs.
2. $\log_a M - \log_a N = \log_a \dfrac{M}{N} \leftarrow$ The Quotient Property of Logs.
3. $(\log_a M)^k = k \log_a M \leftarrow$ The Power Property of Logs.

We use the properties to solve exponential equations.

$$3^x = 4^{2x-1}$$

We see we can't get like bases. Additionally, since we have two different exponential bases, we can't switch to a single logarithmic form. Let's take the natural log of both side.

$$\ln 3^x = \ln 4^{2x-1} \rightarrow$$

Now use the Power Property of Logs

$$x \ln 3 = (2x - 1) \ln 4$$

Distribute and solve for x

$$x \ln 3 = 2x \ln 4 - \ln 4$$

$$x \ln 3 - 2x \ln 4 = -ln4$$

$$x(\ln 3 - 2 \ln 4) = -\ln 4$$

$$x = \frac{-\ln 4}{\ln 3 - 2 \ln 4} \leftarrow \text{And that is our beautiful answer!}$$

We need a few more basic properties of logarithms to solve our logarithmic equations.

Some Additional Properties of Logs

1. $\log_a a = 1$, and in particular, $\ln e = 1$
2. $\log_a 1 = 0$
3. $\log_a a^x = x$
4. $a^{\log_a x} = x$

Logarithms, like exponentials, have the uniqueness property. That is

$$\log_a M = \log_a N \rightarrow M = N$$

We use all of the properties to solve equations that contain logarithms.

Solve

$$\log_2(x - 1) + \log_2 x = \log_2(x^2 - 3)$$

Use the properties to combine the logs and then use the uniqueness property.

$$\log_2(x - 1) + \log_2 x = \log_2(x - 1)x = \log_2(x^2 - 3)$$

$$x(x - 1) = x^2 - 3$$

$$x^2 - x = x^2 - 3$$

$$-x = -3 \rightarrow x = 3$$

Another example

$$\ln x - \ln(x - 1) = 2$$

$$\ln \frac{x}{x - 1} = 2$$

Switch to exponential form

$$e^2 = \frac{x}{x - 1}$$

$$e^2(x - 1) = x$$

$$e^2 x - e^2 = x$$

$$e^2 x - x = e^2$$

$$x(e^2 - 1) = e^2$$

$$x = \frac{e^2}{e^2 - 1}$$

Some important things to keep in mind.

1. $\log_a(NEGATIVE \text{ or } ZERO)$ is undefined.

2. $\log(M + N) \neq \log M + \log N$

3. $\log x = \log_{10} x$ and is called the common log.

4. $\ln x = \log_e x$ and is called the exponential log.

Practice Exponential Equations

1. $3^{2x-1} = 27$

2. $\left(\dfrac{1}{4}\right)^{-2} = 16^{-1}$

3. $e^x + 6 = 3e^x$

4. $2^x = 3^{x+1}$

5. $4e^{1-x} = 7$

6. $2^x 4^{-x} = 16$

Practice Logarithmic Equations

1. $\log_2(3 - x) = 3$

2. $2 + \log(x - 2) = 4$

3. $\log_2(x^2 + 3x - 1) = 3$

4. $\log_3(x + 1) - \log_3(2x + 1) = 2$

5. $\log_2(\log_3(x - 2)) = 2$

6. $\ln(1 - x) + \ln(2x + 1) = 0$

7. $3^{3\log_3 x} = 7$

8. $\ln(e^2 x + 3e + e) = \ln(2x) + 1$

9. $\log_3(x + 1) + \log_2 8 = \log_3(3 - x) + \log_5 25$

K. Trigonometric Equations

Know.Your.Unit.Circle

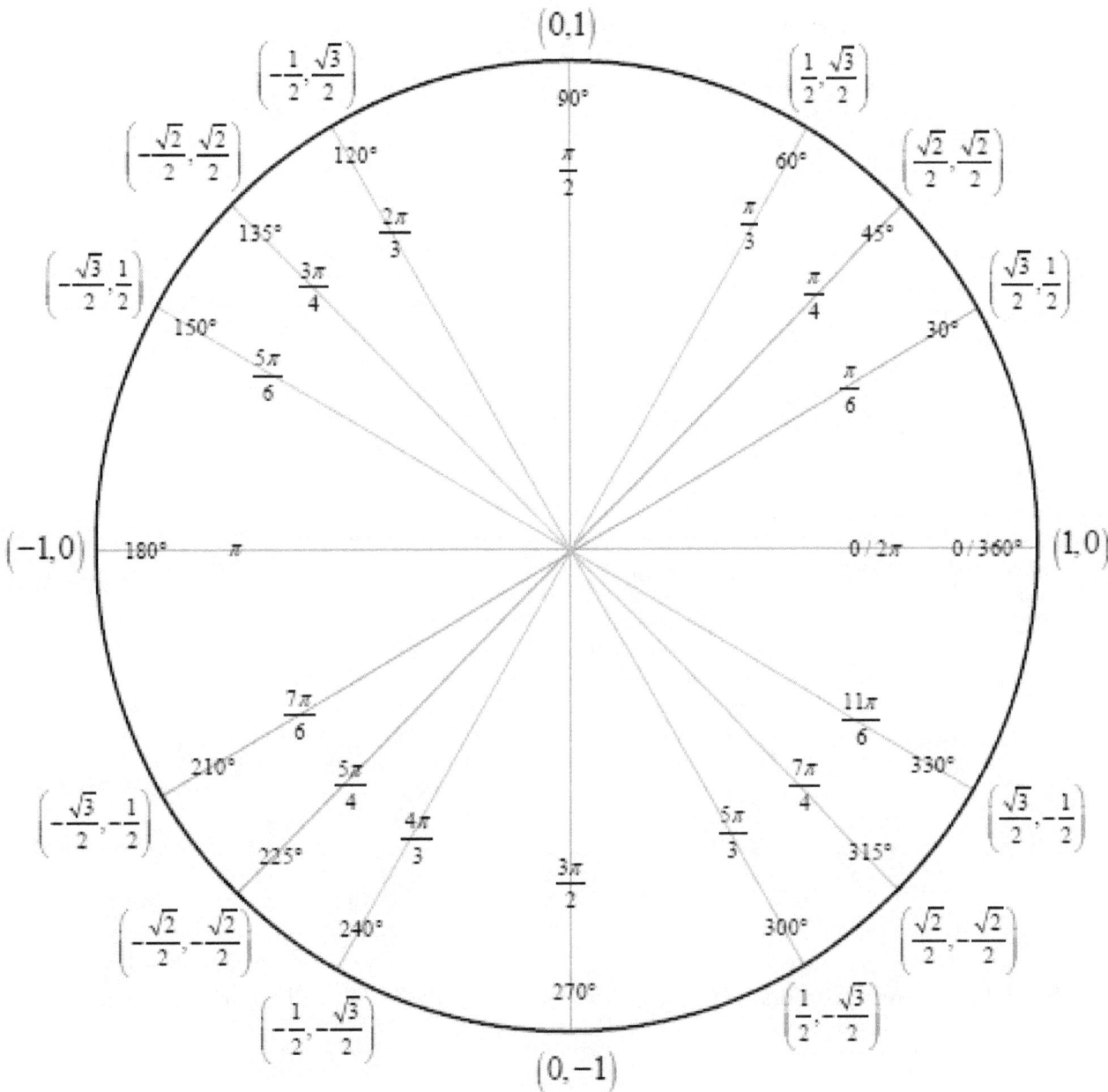

The three Pythagorean Identities:

1. $\sin^2 x + \cos^2 x = 1$
2. $\tan^2 x + 1 = \sec^2 x$
3. $1 + \cot^2 x = \csc^2 x$

The Double Angle Identities for sine and cosine:

1. $\sin(2x) = 2\sin x \cos x$
2. $\cos(2x) = 2\cos^2 x - 1 = 1 - 2\sin^2 x = \cos^2 x - \sin^2 x$

Since our trigonometric functions are periodic, there are often numerous solutions to our equations. You must pay attention to the domain imposed upon the equation to know what values are sought.

For example, solve for all x on $[0, 2\pi)$

$$2\sin x - 1 = 0$$

$$2\sin x = 1$$

$$\sin x = \frac{1}{2}$$

We can see from the unit circle, that there are two values on $[0, 2\pi)$ for which $\sin x = \frac{1}{2}$

$$x = \frac{\pi}{6} \text{ and } x = \frac{5\pi}{6}$$

If, instead, you are asked to solve a trig equation without any restrictions on the domain, then there may be infinitely many solutions.

$$2\cos x + \sqrt{3} = 0$$

$$2\cos x = -\sqrt{3}$$

$$\cos x = -\frac{\sqrt{3}}{2}$$

One suggestion is to find the solutions in just $[0, 2\pi)$ and then add integer multiples of 2π to each.

So in $[0, 2\pi)$, our solutions are $x = \frac{\pi}{3}$ and $x = \frac{5\pi}{3}$. Now to find all solutions, we get

$$x = \frac{\pi}{3} + 2n\pi \text{ and } x = \frac{5\pi}{3} + 2n\pi \text{ for all integers } n.$$

Some things to keep in mind when solving trigonometric equations:

1. You will often use the identities to solve trigonometric equations.

2. Often a good technique is switch everything into cosine and sine.

3. The Pythagorean identities are your friends.

Practice Trigonometric Equations

1. $\cos x \sin x = \sin x$

2. $4\sin^2 x - 1 = 0$

3. $(\tan x - \sqrt{3})(\sec x + 2) = 0$

4. $\csc^2 x - 2 = 0$

5. $\cot x - \sqrt{3} = 0$

6. $2\cos^2 x = \sin x + 1$

7. $3 \sin 2x - 2 \sin x = 0$

8. $\tan^2 x - 2 \sec x = 2$

9. $\cot^2 x + \csc x = 0$

10. $\sin^2 + 3 = 7 \cos^2 x$

L. Things that Look A LOT Like Quadratics

Some Notes:

We visited solving different types of quadratic polynomial in previous sections. The new feature of this section is that sometimes we have equations that model quadratics and can be solved in the same ways. The trick is to recognize them.

Consider the equation

$$(x-2)^2 + 2(x-2) - 8 = 0$$

If we do a substitution $u = x - 2$ then our equation becomes

$$u^2 + 2u - 8 = 0$$

. Now we see that this equation factors

$$(u+4)(u-2) = 0$$

$$u + 4 = 0 \rightarrow u = -4 \quad \text{OR}$$

$$u - 2 = 0 \rightarrow u = 2$$

Since we are interested in solving for x, we "un"-substitute

$$u = x - 2 = -4 \rightarrow x = -2 \quad \text{OR}$$

$$u = x - 2 = 2 \rightarrow x = 4$$

The hardest part will be recognizing the equation as a one that could be modeled as a quadratic. Here are some common ones:

(a) $\cos^2 x + \cos x - 2 = 1; u = \cos x$

(b) $3a^{2x} - 2a^x + 1 = 0; u = a^x$

(c) $x + x^{1/2} + 4 = 1; u = x^{1/2}$

With the substitution indicated, they all become

(a) $u^2 + u - 2 = 1$

(b) $3u^2 - 2u + 1 = 0$

(c) $u^2 + u + 4 = 1$

Easy Peasy.

Practice Things that Look A LOT Like Quadratics

1. $(x+1)^2 - 5(x+1) - 6$

2. $2e^{2x} + 3e^x - 2$

3. $\sin^2(x) + \sin(x) = 2$

4. $\sin^2 x = 2\sin x + 3$

5. $x^{1/3} + x^{1/6} - 2 = 0$

M. Inequalities

Some Notes:

Here we diverge from our journey with equations and tackle solving inequalities. Many similar techniques are used but there is often an extra layer of difficulty.

Unlike with equations, the ZPP does not apply!

Recall with equations, if $a \cdot b = 0$ then $a = 0$ OR $b = 0$

But with inequalities, if $a \cdot b > 0$ then we cannot conclude that $a > 0$ or $b > 0$. In fact if a and b are both negtive, then their product is positive.

Often a sign chart is a useful way to solve inequalities that can be written as a product of factors.

Our goal is to determine where each factor is positive and negative. Note that each factor will switch signs at its zero. For example, given a factor of $(x + 2)$, we see that when $x = -2$ is its zero. Furthermore, we see that $x + 2 < 0$ when $x < -2$ and that $x + 2 > 0$ when $x > -2$. Let's apply this the following example.

$$(x + 3)(x - 1) < 0$$

Factors	$x < -3$	$-3 < x < 1$	$x > 1$
$x + 3$	-	+	+
$x - 1$	-	-	+
$(x + 3)(x - 1)$	+	-	+

We see from the sign chart that our solution set is $(-3, 1)$

Some things to keep in mind when working with inequalities to remember

1. When multiplying both sides of an inequality by a negative number, the inequality flips directions. That is, $a < b \rightarrow -a > -b$

2. If $a < b$, then $\dfrac{1}{a} > \dfrac{1}{b}$

Absolute value inequalities add a layer still.

To solve $|x| < a$, convert the inequality to $-a < x < a$

To solve $|x| > a$, convert the inequality to $x > a \quad$ OR $\quad x < -a$

Practice Inequalities

1. $\dfrac{3}{x} < \dfrac{1}{2}$

2. $2 \leq 2x - 1 < 3$

3. $x^2 - 2x > 0$

4. $(x + 2)(x - 3) < -4$

5. $\dfrac{2 - x}{x + 1} > 4$

6. $\dfrac{(2 - x)(x + 1)}{3 - x} < 0$

7. $|2x - 1| < 3$

8. $|3 - x| > 2$

9. $x^2 - 3x - 10 \geq 0$

10. $2x^2 + 3x - 3 < -1$

N. Finding Equations of LInes

Some Notes:

Lines are easy and finding them happens every calculus day. Recall some of our formulas

1. $y = mx + b \leftarrow$ slope-intercept formula of a line (probably the most familiar)
2. $(y - y_1) = m(x - x_1) \leftarrow$ point-slope formula of a line (the most used in calculus)
3. $m = \dfrac{y_2 - y_1}{x_2 - x_1} \leftarrow$ slope formula

Some more fun facts to remember about lines:

1. Parallel lines have the same slope
2. Perpendicular lines have slopes that are negative reciprocals of each other
3. To find the slope of a line in standard form, $ax + by = c$, solve for y and get the line in slope-intercept form.

Bonus Thought: The equation of a circle centered at (h, k) with radius r is

$$(x - h)^2 + (y - k)^2 = r^2$$

Now.Go.Practice.

Practice Finding Equations of Lines

1. Find the equation of the line passing through the point $(2, 3)$ with slope $m = -2$

2. Find the equation of the line passing through the points $(4, 5)$ and $(-2, 7)$

3. Find the equation of the line passing through $(-2, 4)$ and parallel to the line $2x - 3y = 7$

4. Find the equation of the line passing through $(3, -1)$ and perpendicular to the line
$3x + 7y = -1$

Bonus

Find the equation of the circle with center $(-2, 3)$ and radius $r = 2$.

Part IV - The Solutions

A. Linear Equations and Absolute Value Equations

1. $4(x-1)-(3-x)=2 \rightarrow$ Distribute

   $4x-4-3+x=2 \rightarrow$ Combine like terms

   $5x-7=2 \rightarrow$ Isolate the x term

   $5x=9$

   $x=\dfrac{9}{5}$

2. $1-[3-[4-[5-x]]]=6(2-x)$

   $1-3+[4-[5-x]]=12-6x$

   $-2+4-5+x=12-6x$

   $-3+x=12-6x$

   $x+6x=12+3$

   $7x=15$

   $x=\dfrac{15}{7}$

3. $-3p+4=-2(1-p)$

   $-3p+4=-2+2p$

   $-3p-2p=-2-4$

   $-5p=-6$

   $p=\dfrac{-6}{-5}=\dfrac{6}{5}$

4. $3x-2-(x-4)=6(-(3-2x))$

   $3x-2-x+4=6(-3+2x)$

   $2x+2=-18+12x$

   $2x-12x=-18-2$

   $-10x=-20$

   $x=2$

5. $|2x| = 4$

$2x = 4$ or $2x = -4$

$x = 2$ or $x = -2$

6. $|3 - (x - 2)| = 6$

$3 - (x - 2) = 6$ or $3 - (x - 2) = -6$

$3 - x + 2 = 6$ or $3 - x + 2 = -6$

$5 - x = 6$ or $5 - x = -6$

$-x = 1$ or $-x = -11$

$x = -1$ or $x = 11$

7. $-2|3(x - 1)| = -4$

$|3(x - 1)| = 2$

$3(x - 1) = 2$ or $3(x - 1) = -2$

$3x - 3 = 2$ or $3x - 3 = -2$

$3x = 5$ or $3x = 1$

$x = \dfrac{5}{3}$ or $x = \dfrac{1}{3}$

8. $|3 - x| = |2x + 1|$

$3 - x = 2x + 1$ or $3 - x = -(2x + 1) = -2x - 1$

$3 - 1 = 2x + x$ or $-x + 2x = -1 - 3$

$2 = 3x$ or $x = -4$

$x = \dfrac{2}{3}$ or $x = -4$

B. Equations with Fractions

1. $\dfrac{1}{2}x + \dfrac{2}{3}x - \dfrac{3}{4}x = 2 \to$ multiply by the LCD

$12\left(\dfrac{1}{2}x + \dfrac{2}{3}x - \dfrac{3}{4}x = 2\right)$

$$12 * \frac{1}{2}x + 12 * \frac{2}{3}x - 12 * \frac{3}{4}x = 12 * 2$$

$$6x + 4 * 2x - 3 * 3x = 24$$

$$6x + 8x - 9x = 24$$

$$5x = 24$$

$$x = \frac{24}{5}$$

2. $\frac{1}{2}(x - 2) + 3x = \frac{3}{2}(2 - x)$. Multiply by the LCD = 2.

$2\left(\frac{1}{2}(x - 2) + 3x = \frac{3}{2}(2 - x)\right)$. Cancel the fractions.

$$(x - 2) + 2(3x) = 3(2 - x)$$

$$x - 2 + 6x = 6 - 3x$$

$$7x = 8 - 3x$$

$$10x = 8$$

$$x = \frac{8}{10} = \frac{4}{5}$$

3. $\frac{2}{3}x - \frac{3}{4}(2 + x) = \frac{1}{6}(x - 1) + x$. Multiply by LCD = 12

$12\left(\frac{2}{3}x - \frac{3}{4}(2 + x) = \frac{1}{6}(x - 1) + x\right)$. Cancel fractions.

$$4(2x) - 3(3(2 + x)) = 2(x - 1) + 12x$$

$$8x - 9(2 + x) = 2x - 2 + 12x$$

$$8x - 18 - 9x = 14x - 2$$

$$-x - 14x = -2 + 18$$

$$-15x = 16$$

$$x = -\frac{16}{15}$$

4. $\dfrac{1 + \frac{1}{x}}{x - \frac{x}{2}} = 3$. Create one fraction in the numerator and the denominator.

$$\frac{\dfrac{x}{x}+\dfrac{1}{x}}{\dfrac{2x}{2}-\dfrac{x}{2}}=3$$

$$\frac{\dfrac{x+1}{x}}{\dfrac{2x-x}{2}}=3$$

$$\frac{x+1}{x}*\frac{2}{x}=3$$

$$\frac{2(x+1)}{x^2}=3$$

$$2x+2=3x^2$$

$$0=3x^2-2x-2$$

$$x=\frac{2\pm\sqrt{4-4(3)(-2)}}{6}$$

$$x=\frac{2\pm\sqrt{28}}{6}=\frac{2\pm2\sqrt{7}}{6}=\frac{1\pm\sqrt{7}}{3}$$

5. $$\frac{\dfrac{1}{2}-\dfrac{3}{4}x}{\dfrac{2}{3x}+\dfrac{5}{6}}=0$$

Find a common denominator on both top and bottom.

$$\frac{\dfrac{2}{4}-\dfrac{3}{4}x}{\dfrac{4}{6x}+\dfrac{5x}{6x}}=0$$

Combine fractions

$$\frac{\dfrac{2-3x}{4}}{\dfrac{4+5x}{6x}}=0$$

Since a fraction is equal to zero only when the numerator is zero, we need only set the numerator equal to zero.

$$\frac{2-3x}{4}=0$$

$$2-3x=0\rightarrow x=\frac{2}{3}$$

C. Solving for a Variable Amongst a Sea of Letters (5)

1. Solve for a; $2a^2 - 3b^2 = c^2$

$$2a^2 = c^2 + 3b^2$$

$$a^2 = \frac{c^2 + 3b^2}{2}$$

$$a = \pm\sqrt{\frac{c^2 + 3b^2}{2}}$$

2. Solve for M; $F = G\frac{mM}{r^2}$

$$\frac{F}{G} = \frac{mM}{r^2}$$

$$\frac{r^2 F}{G} = mM$$

$$\frac{r^2 F}{mG} = M$$

3. Solve for h; $S = 2\pi r^2 + 2\pi rh$

$$S - 2\pi r^2 = 2\pi rh$$

$$\frac{S - 2\pi r^2}{2\pi r} = h$$

Is it OK to cancel the $2\pi r$, leaving

$S - r$? NO!!! Remember, we cannot cancel across a sum unless you divide into each term.

4. Solve for y; $x = \frac{1 - y}{y}$

$$xy = 1 - y$$

$$xy + y = 1$$

$$y(x + 1) = 1$$

$$y = \frac{1}{x + 1}$$

5. Solve for d; $2acd + 3(c - d) = a + 2(b - d)$

$$2acd + 3c - 3d = a + 2b - 2d$$

$$2acd - 3d + 2d = a + 2b - 3c$$

72

$$2acd - d = d(2ac - 1) = a + 2b - 3c$$

$$d = \frac{a + 2b - 3c}{2ac - 1}$$

6. Solve for $\frac{dy}{dx}$; $3x\frac{dy}{dx} - 2\left(1 - \frac{dy}{dx}\right) = 3 + x\frac{dy}{dx}$

$$3x\frac{dy}{dx} - 2 + 2\frac{dy}{dx} = 3 + x\frac{dy}{dx}$$

$$3x\frac{dy}{dx} + 2\frac{dy}{dx} - x\frac{dy}{dx} = 3 + 2 = 5$$

$$(3x + 2 - x)\frac{dy}{dx} = 5$$

$$(2x + 2)\frac{dy}{dx} = 5$$

$$\frac{dy}{dx} = \frac{5}{2x + 2}$$

7. Solve for l; $\frac{1 + 3l}{l - 1} = \pi$

$1 + 3l = \pi(l - 1) = \pi l - \pi$ Group the π terms.

$$1 + 3l - \pi l = -\pi$$

$$3l - \pi l = -\pi - 1$$

$$l(3 - \pi) = -\pi - 1$$

$$l = \frac{-\pi - 1}{3 - \pi}$$

D. Basic factoring

1. $x^8 - 1 = 0$

$$(x^4 + 1)(x^4 - 1) = 0$$

$$(x^4 + 1)(x^2 + 1)(x^2 - 1) = 0$$

$(x^4 + 1)(x^2 + 1)(x + 1)(x - 1) = 0$ Now use the ZPP

$x^4 + 1 = 0$ or $x^2 + 1 = 0$ or $x + 1 = 0$ or $x - 1 = 0$

We see that $x^4 + 1 \neq 0$, $x^2 + 1 \neq 0$ but $x = -1$ when $x + 1 = 0$ and $x = 1$ when $x - 1 = 0$.

2. $4x^2 - 36 = 0$

   $4(x^2 - 9) = 0$

   $x^2 - 9 = 0$

   $(x - 3)(x + 3) = 0$ ZPP

   $x - 3 = 0 \rightarrow x = 3$

   $x + 3 = 0 \rightarrow x = -3$

3. $x^2 + 1 = 0$

   $x^2 = -1$ Oh no!

   We see that there is no real x value that when squared yields -1. Thus there is no real solution to this equation.

4. $(x + 2)^4 - (x + 2)^2 = 0$

   $(x + 2)^2[(x + 2)^2 - 1] = 0$ ZPP!

   $(x + 2)^2 = 0$ or $(x + 2)^2 - 1 = 0$

   When $(x + 2)^2 = 0 \rightarrow x + 2 = 0 \rightarrow x = -2$

   $(x + 2)^2 - 1 = 0 \rightarrow (x + 2)^2 = 1 \rightarrow x + 2 = \pm 1 \rightarrow x = -2 \pm 1$

   $x = -2 + 1 = -1$ or $x = -2 - 1 = -3$

5. $x^6 - 9 = -1$

   $x^6 - 8 = 0$ This is the difference of two cubes.

   $(x^2 - 2)(x^4 + 2x^2 + 4) = 0$ ZPP

   $x^2 - 2 = 0 \rightarrow x^2 = 2 \rightarrow x = \pm\sqrt{2}$ OR

   $x^4 + 2x^2 + 4 = 0$ which has no real solution.

6. Solve for all x and y that satisfy this equation: $6xy^2 - 9xy - 6x = 0$

   $3x(2y^2 - 3y - 2) = 0$

   $3x(2y + 1)(y - 2) = 0$ ZPP

   $3x = 0 \rightarrow x = 0$ OR

   $2y + 1 = 0 \rightarrow y = -\dfrac{1}{2}$ OR

   $y - 2 = 0 \rightarrow y = 2$

E. Quadratic Equations by Factoring

1. $2x^2 + 3x - 2 = 0$

$(2x - 1)(x + 2) = 0$

$2x - 1 = 0 \rightarrow x = \dfrac{1}{2}$ OR

$x + 2 = 0 \rightarrow x = -2$

2. $-3x^2 + x = -2$

$-3x^2 + x + 2 = 0$

$(-3x - 2)(x - 1) = 0$

$-3x - 2 = 0 \rightarrow -3x = 2 \rightarrow x = -\dfrac{2}{3}$ OR

$x - 1 = 0 \rightarrow x = 1$

3. $x^2 - 7 = 0$

$(x - \sqrt{7})(x + \sqrt{7}) = 0$

$x - \sqrt{7} = 0 \rightarrow x = \sqrt{7}$ OR

$x + \sqrt{7} = 0 \rightarrow x = -\sqrt{7}$

4. $6x^2 - x = 2$

$(3x - 2)(2x + 1) = 0$

$3x - 2 = 0 \rightarrow 3x = 2 \rightarrow x = \dfrac{2}{3}$ OR

$2x + 1 = 0 \rightarrow 2x = -1 \rightarrow -\dfrac{1}{2}$

5. $-2x^2 + 10x = 12$

$2(x - 2)(-x + 3) = 0$

$x - 2 = 0 \rightarrow x = 2$ OR

$-x + 3 = 0 \rightarrow x = 3$

F. By the Quadratic Formula

1. $ax^2 + bx + c = 0$

$$x = \frac{-b \pm \sqrt{b^2 - 4ac}}{2a}$$

2. $3x^2 - 4x = \dfrac{1}{2}$

$$3x^2 - 4x - \frac{1}{2} = 0$$

$$x = \frac{4 \pm \sqrt{16 - 4(3)(-1/2)}}{2 * 3}$$

$$x = \frac{4 \pm \sqrt{16 + 6}}{6}$$

$$x = \frac{4 \pm \sqrt{22}}{6}$$

3. $2x - 7 = -x - x^2$

First get your quadratic into standard form.

$$x^2 + x + 2x - 7 = x^2 + 3x - 7 = 0$$

$$x = \frac{-3 \pm \sqrt{9 - 4(1)(-7)}}{2}$$

$$x = \frac{-3 \pm \sqrt{37}}{2}$$

4. $px - qx^2 = sx^2 + 2$

$$-sx^2 - qx^2 + px - 2 = 0$$

$(-s - q)x^2 + px - 2 = 0$. Note that $a = -s - q, b = p, c = -2$

$$x = \frac{-p \pm \sqrt{p^2 - 4(-s - q)(-2)}}{2(-s - q)} \qquad \text{Simplify}$$

$$x = \frac{-p \pm \sqrt{p^2 + 8(-s - q)}}{2(-s - q)}$$

5. $\frac{1}{2}x^2 - \frac{2}{3}x - 2 = 0$. I see fractions - get rid of 'em! Multiply by LCD = 6.

$$6\left(\frac{1}{2}x^2 - \frac{2}{3}x - 2 = 0\right)$$

$$3x^2 - 4x - 12 = 0$$

$$x = \frac{4 \pm \sqrt{16 - 4(3)(-12)}}{6}$$

$$x = \frac{4 \pm \sqrt{16 + 144}}{6}$$

$$x = \frac{4 \pm \sqrt{160}}{6} = \frac{4 \pm 4\sqrt{10}}{6} = \frac{2 \pm 2\sqrt{10}}{3}$$

G. By Completing the Square

1. $x^2 + 4x - 7 = 0$

$x^2 + 4x = 7$

$x^2 + 4x + 4 = 7 + 4$

$(x + 2)^2 = 11$

$x + 2 = \pm\sqrt{11}$

$x = -2 \pm \sqrt{11}$

2. $3x^2 - 6x = 9$

Divide first by 3

$x^2 - 2x = 3$

$x^2 - 2x + 1 = 4$

$(x - 1)^2 = 4$

$x - 1 = \pm 2$

$x = 1 \pm 2$

$x = 1 + 2 = 3$ OR

$x = 1 - 2 = -1$

3. $2x^2 - 3x = 5$

Divide first by 2

$$x^2 - \frac{3}{2}x = \frac{5}{2}$$

$$x^2 - \frac{3}{2} + \frac{9}{16} = \frac{5}{2} + \frac{9}{16} = \frac{40}{16} + \frac{9}{16} = \frac{49}{16}$$

$$\left(x - \frac{3}{4}\right)^2 = \frac{49}{16}$$

$$x - \frac{3}{4} = \pm\frac{7}{4}$$

$$x = \frac{3}{4} \pm \frac{7}{4}$$

$$x = \frac{3}{4} + \frac{7}{4} = \frac{10}{4} = \frac{5}{2} \text{ OR}$$

$$x = \frac{3}{4} - \frac{7}{4} = \frac{-4}{4} = -1$$

4. $\frac{1}{3}x^2 - 2x + 1 = 2$

Multiply by LCD = 3

$$3\left(\frac{1}{3}x^2 - 2x + 1 = 2\right)$$

$x^2 - 6x + 3 = 6$. Now complete the square.

$x^2 - 6x + 9 + 3 = 6 + 9 = 15$

$x^2 - 6x + 9 = 12$

$(x-3)^2 = 12$

$x - 3 = \pm\sqrt{12}$

$x = 3 \pm \sqrt{12}$

$x = 3 \pm 2\sqrt{3}$

5. Solve for x; $ax^2 + bx + c = 0$

$$\frac{a}{a}x^2 + \frac{b}{a}x + \frac{c}{a} = \frac{0}{a} = 0$$

$$x^2 + \frac{b}{a}x = -\frac{c}{a}$$

$$x^2 + \frac{b}{a}x + \left(\frac{1}{2}\frac{b}{a}\right)^2 = -\frac{c}{a} + \left(\frac{1}{2}\frac{b}{a}\right)^2$$

$$x^2 + \frac{b}{a}x + \frac{b^2}{4a^2} = \frac{b^2}{4a^2} - \frac{c}{a} = \frac{b^2}{4a^2} - \frac{4ac}{4a^2} = \frac{b^2 - 4ac}{4a^2}$$

$$\left(x + \frac{b}{2a}\right)^2 = \frac{b^2 - 4ac}{4a^2}$$

$$x + \frac{b}{2a} = \pm\sqrt{\frac{b^2 - 4ac}{4a^2}} = \pm\frac{\sqrt{b^2 - 4ac}}{2a}$$

$$x = -\frac{b}{2a} \pm \frac{\sqrt{b^2 - 4ac}}{2a}$$

$$x = \frac{-b \pm \sqrt{b^2 - 4ac}}{2a} \quad \leftarrow \text{How cool is that?}$$

H. Equations with Fractional or Negative Exponents

1. $4(x-1)^{1/2} - 3(x-1)^{3/2} + (x-1)^{5/2} = 0$

   Factor out the common factor with the lowest power, $(x-1)^{1/2}$

   $(x-1)^{1/2}(4 - 3(x-1)^{2/2} + (x-1)^{4/2}) = 0$

   $(x-1)^{1/2}(4 - 3(x-1) + (x-1)^2) = 0$ Simplify

   $(x-1)^{1/2}(4 - 3x + 3 + x^2 - 2x + 1) = 0$

   $(x-1)^{1/2}(x^2 - 5x + 8) = 0 \qquad$ Now use the ZPP

   $(x-1)^{1/2} = 0$ OR $x^2 - 5x + 8 = 0$

   When $(x-1)^{1/2} = 0, \rightarrow x = 1$

   When $x^2 - 5x + 8 = 0$, use the quadratic formula

   $x = \dfrac{5 \pm \sqrt{25 - 4(8)}}{2}$. We see immediately this does not yield a real solution.

   So our final answer is $x = 1$.

2. $x^{1/2} + 3x^{-1/2} = 10x^{-3/2}$

   $x^{1/2} + 3x^{-1/2} - 10x^{-3/2} = 0$

   Factor out the common factor with the lowest power, $x^{-3/2}$

   $x^{-3/2}(x^{4/2} + 3x^{2/2} - 10) = 0$

   $x^{-3/2}(x^2 + 3x - 10) = 0$

   $x^{-3/2}(x + 5)(x - 2) = 0$

$$\frac{(x+5)(x-2)}{x^{3/2}} = 0 \qquad \text{Use the ZPP}$$

$$x + 5 = 0 \rightarrow x = -5 \text{ OR } x - 2 = 0 \rightarrow x = 2$$

3. $(x-1)^{1/2} + 3x(x-1)^{-1/2} = 0$

$(x-1)^{-1/2}((x-1) + 3x) = 0$

$$\frac{4x-1}{(x-1)^{1/2}} = 0$$

$$4x - 1 = 0 \rightarrow x = \frac{1}{4}$$

4. $(x+2)^{4/5} - (2x+4)^{-1/5} = 0$

$(x+2)^{4/5} - (2(x+2))^{-1/5} = 0$

$(x+2)^{4/5} - 2^{-1/5}(x+2)^{-1/5} = 0$

$(x+2)^{-1/5}((x+2)^{5/5} - 2^{-1/5}) = 0$

$$\frac{x+2 - 2^{1/5}}{(x+2)^{1/5}} = 0$$

$$x + 2 - 2^{-1/5} = 0 \rightarrow x = 2^{-1/5} - 2$$

5. $3(x+1)^{-1} - 3(x+1)^{-2} = -2$

$3(x+1)^{-2}((x+1) - 1) = -2$

$$\frac{3x}{(x+1)^2} = -2$$

$$3x = -2(x^2 + 2x + 1) = -2x^2 - 4x - 2 \qquad \text{A quadratic!}$$

$$2x^2 + 7x + 2 = 0$$

$$x = \frac{-7 \pm \sqrt{49 - 4(2)(2)}}{4}$$

$$x = \frac{-7 \pm \sqrt{33}}{4}$$

I. Radical Equations

1. $\sqrt{x-2} = 4$

$(\sqrt{x-2})^2 = 4^2$

$x - 2 = 16 \rightarrow x = 18$

Check: $\sqrt{18 - 2} = \sqrt{16} = 4$ All good!

2. $3\sqrt{x+1} = (3-x)\sqrt{x+1}$

$3\sqrt{x+1} = 3\sqrt{x+1} - x\sqrt{x+1}$

$0 = -x\sqrt{x+1}$

$x = 0$ OR $\sqrt{x+1} = 0 \rightarrow x + 1 = 0 \rightarrow x = -1$

3. $\sqrt{3-x} + 2 = \sqrt{x+1}$

$(\sqrt{3-x} + 2)^2 = (\sqrt{x+1})^2$

$(3-x) + 2\sqrt{3-x} + 2\sqrt{3-x} + 4 = x + 1$

$7 - x + 4\sqrt{3-x} = x + 1$ Isolate the radical

$4\sqrt{3-x} = 2x - 6$

$\sqrt{3-x} = \dfrac{2x-6}{4} = \dfrac{x-3}{2}$. Squaring both sides,

$3 - x = \dfrac{(x-3)^2}{2^2} = \dfrac{x^2 - 6x + 9}{4}$

$4(3-x) = 12 - 4x = x^2 - 6x + 9$ A quadratic! Set it to zero

$0 = x^2 - 2x - 3$

$0 = (x-3)(x+1)$

$0 = x - 3 \rightarrow x = 3$ OR $0 = x + 1 \rightarrow x = -1$ Now go back and check!

For $x = 3$,

$\sqrt{3-3} + 2 = \sqrt{0} + 2 = 0 + 2 = 2$ and $\sqrt{3+1} = \sqrt{4} = 2$. Thus, $x = 3$ is a solution.

For $x = -1$,

$\sqrt{3--1} + 2 = \sqrt{4} + 2 = 2 + 2 = 4$ and $\sqrt{-1+1} = 0$. Oh no! Thus, $x = -1$ is extraneous and NOT a solution.

4. $\sqrt{\sqrt{x}+5} - 5 = 0$

$\sqrt{\sqrt{x}+5} = 5$

$(\sqrt{\sqrt{x}+5})^2 = 5^2 = 25$

$\sqrt{x} + 5 = 25$

$\sqrt{x} = 20$

$x = 400$. Make sure to check!

J. Rational Equations

1. $\dfrac{5}{x} + \dfrac{3}{x-2} = 4$

Mutiply through by LCD $= x(x-2)$

$\dfrac{x(x-2)}{1}\left(\dfrac{5}{x} + \dfrac{3}{x-2} = 4\right)$

$(x-2)5 + 3x = 4x(x-2)$

$5x - 10 + 3x = 4x^2 - 8x$

$4x^2 - 16x + 10 = 0$ Divide by 2

$2x^2 - 8x + 5 = 0$

$x = \dfrac{8 \pm \sqrt{64 - 4(2)(5)}}{4}$

$x = \dfrac{8 \pm \sqrt{24}}{4} = \dfrac{8 \pm 2\sqrt{6}}{4} = \dfrac{4 \pm \sqrt{6}}{2}$

2. $\dfrac{1}{x-1} + \dfrac{x+2}{(x-1)^2} = \dfrac{1}{x}$

Multiply by LCD $= x(x-1)^2$

$\dfrac{x(x-1)^2}{1}\left(\dfrac{1}{x-1} + \dfrac{x+2}{(x-1)^2} = \dfrac{1}{x}\right)$

$x(x-1)1 + x(x+2) = (x-1)^2$

$x^2 - x + x^2 + 2x = x^2 - 2x + 1$

$x^2 + 3x - 1 = 0$

$x = \dfrac{-3 \pm \sqrt{9 - 4(-1)}}{2}$

$x = \dfrac{-3 \pm \sqrt{13}}{2}$

3. $\dfrac{3x^2 - 5x - 2}{x^2 - 4} = \dfrac{2}{3}$

Multiply by the LCD $= 3(x^2 - 4)$

82

$$\frac{3(x^2-4)}{1}\left(\frac{3x^2-5x-2}{x^2-4}=\frac{2}{3}\right)$$

$$3(3x^2-5x-2)=2(x^2-4)$$

$$9x^2-15x-6=2x^2-8$$

$$7x^2-15x+2=0$$

$$x=\frac{15\pm\sqrt{225-4(7)(2)}}{14}$$

$$x=\frac{15\pm\sqrt{169}}{14}=\frac{15\pm13}{14}$$

$$x=\frac{28}{14}=2 \text{ OR } x=\frac{2}{14}=\frac{1}{7}$$

4. $$\frac{x+1}{-x^2+5x-6}\cdot\frac{x-2}{x^2+3x+2}=7$$

$$\frac{x+1}{(x-2)(3-x)}\cdot\frac{x-2}{(x+1)(2+x)}=7$$

$$\frac{\cancel{x+1}}{(\cancel{x-2})(3-x)}\cdot\frac{\cancel{x-2}}{(\cancel{x+1})(2+x)}=7$$

$$\frac{1}{(3-x)}\cdot\frac{1}{(2+x)}=7$$

$$1=7(3-x)(2+x)$$

$$1=7(6+x-x^2). \qquad \text{A quadratic!}$$

$$1=42+7x-7x^2$$

$$7x^2-7x-41=0$$

$$x=\frac{7\pm\sqrt{49-4(7)(-41)}}{2}$$

$$x=\frac{7\pm\sqrt{49+1148}}{2}=\frac{7\pm\sqrt{1197}}{2}$$

5. $$\frac{x}{3x-1}\div\frac{x^3}{9x^2-9x+2}=1$$

$$\frac{x}{3x-1}\div\frac{x^3}{(3x-1)(3x-2)}=1$$

$$\frac{x}{3x-1}\cdot\frac{(3x-1)(3x-2)}{x^3}=1$$

$$\frac{\cancel{x}}{\cancel{3x-1}} \cdot \frac{\cancel{(3x-1)}(3x-2)}{x^{\cancel{2}}} = 1$$

$$\frac{3x-2}{x^2} = 1$$

$$3x - 2 = x^2$$

$$x^2 - 3x + 2 = 0$$

$$(x-2)(x-1) = 0$$

$$x = 2 \text{ OR } x = 1$$

K. Exponential Equations

(a) $3^{2x-1} = 27$

Get like bases

$$3^{2x-1} = 3^3$$

Since the bases are the same, use the uniqueness property and set the exponents equal

$$2x - 1 = 3$$

$$2x = 4$$

$$x = 2$$

(b) $\left(\frac{1}{4}\right)^{-2} = 16^{-x}$

$$(4^{-1})^{-2} = (4^2)^{-x}$$

$$4^2 = 4^{-2x}$$

$$2 = -2x$$

$$-1 = x$$

(c) $e^x + 6 = 3e^x$

$$-2e^x = -6$$

$$e^x = 3$$

$$\ln e^x = \ln 3$$

$$x = \ln 3$$

(d) $2^x = 3^{x+1}$

$\ln 2^x = \ln 3^{x+1}$

$x \ln 2 = (x+1) \ln 3 = x \ln 3 + \ln 3$

$x \ln 2 - x \ln 3 = \ln 3$

$x(\ln 2 - \ln 3) = \ln 3$

$x = \dfrac{\ln 3}{\ln 2 - \ln 3}$

(e) $4e^{1-x} = 7$

Isolate the exponent first.

$e^{1-x} = \dfrac{7}{4}$

$\ln e^{1-x} = \ln \dfrac{7}{4}$

$1 - x = \ln \dfrac{7}{4}$

$-x = -1 + \ln \dfrac{7}{4}$

$x = 1 - \ln \dfrac{7}{4}$

(f) $2^x 4^{-x} = 16$

Work to get like bases

$2^x (2^2)^{-x} = 2^4$

$2^x 2^{-2x} = 2^{x-2x} = 2^4$

$2^{-x} = 2^4 \rightarrow -x = 4 \rightarrow x = -4$

L. Logarithmic Equations

1. $\log_2(3 - x) = 3$

Switch to exponential form

$2^3 = 3 - x$

$8 - 3 = -x$

$5 = -x$ so $x = -5.$

2. $2 + \log(x - 2) = 4$

Isolate the logarithm

$\log(x - 2) = 2$. Switch to exponential form.

$10^2 = x - 2$

$100 + 2 = x$

$x = 102$

3. $\log_2(x^2 + 3x - 1) = 3$

$x^2 + 3x - 1 = 2^3 = 8$

$x^2 + 3x - 9 = 0$

$x = \dfrac{-3 \pm \sqrt{9 - 4(-9)}}{2}$

$x = \dfrac{-3 \pm \sqrt{45}}{2}$

4. $\log_3(x + 1) - \log_3(2x + 1) = 2$

Combine the LHS into one log.

$\log_3\left(\dfrac{x + 1}{2x + 1}\right) = 2$. Switch to exponential form

$\dfrac{x + 1}{2x + 1} = 3^2 = 9$

$x + 1 = 9(2x + 1) = 18x + 9$

$-8 = 17x$

$x = -\dfrac{17}{8}.$

However, upon checking the solution, we see this is outside the domain of our equation. Thus, there is no real solution to this problem.

5. $\log_2(\log_3(x - 2)) = 2$

Switch to the exponential form.

$\log_3(x - 2) = 2^2 = 4$

Switch again.

$x - 2 = 3^4 = 81$

$x = 83$

6. $\ln(1 - x) + \ln(2x + 1) = 0$

$\ln(1 - x)(2x + 1) = 0$

$(1 - x)(2x + 1) = e^0 = 1$

$2x + 1 - 2x^2 - x = 1$

$-2x^2 + x = 0$

$-x(2x - 1) = 0$

$x = 0 \text{ OR } x = \dfrac{1}{2}.$

7. $3^{3\log_3 x} = 7$

Use the power property of logs

$3^{\log_3 x^3} = 7$

Recognize this is the composition of inverse functions

$x^3 = 7$

$x = 7^{1/3}$

8. $\ln(e^2 x + 3e + e) = \ln(2x) + 1$

$\ln(e(ex + 3 + 1) = \ln(2x) + 1$

$\ln e + \ln(ex + 4) = \ln(2x) + 1$

$1 + \ln(ex + 4) = \ln(2x) + 1$

$\ln(ex + 4) = \ln(2x)$

$ex + 4 = 2x$

$ex - 2x = -4$

$x(e - 2) = -4$

$x = -\dfrac{4}{e - 2} = \dfrac{4}{2 - e}$

9. $\log_3(x + 1) + \log_2 8 = \log_3(3 - x) + \log_5 25$

$\log_3(x + 1) + 3 = \log_3(3 - x) + 2$

$$\log_3(x+1) - \log_3(3-x) = -1$$

$$\log_3 \frac{x+1}{3-x} = -1$$

$$\frac{x+1}{3-x} = 3^{-1} = \frac{1}{3}$$

$$3(x+1) = 3 - x$$

$$3x + 3 = 3 - x$$

$$4x = 0$$

$$x = 0$$

M. Trigonometric Equations

1. $\cos x \sin x = \sin x$

$\cos x \sin x - \sin x = 0$

$\sin x (\cos x - 1) = 0$ Use the famed ZPP!

$\sin x = 0$ or $\cos x - 1 = 0$

When $\sin x = 0$, $x = n\pi$, for all integers n.

When $\cos x - 1 = 0 \rightarrow \cos x = 1$, then $x = 2n\pi$ for all integers n.

2. $4\sin^2 x - 1 = 0$

$4\sin^2 x = 1$

$\sin^2 x = \frac{1}{4}$

$\sin x = \pm\sqrt{\frac{1}{4}} = \pm\frac{1}{2}$

$x = \frac{\pi}{6} + n\pi$ OR $x = \frac{5\pi}{6} + n\pi$ for all integers n.

3. $\tan x \sec x + 2\tan x - \sqrt{3}\sec x - 2\sqrt{3}$

$(\tan x - \sqrt{3})(\sec x + 2) = 0$

$\tan x = \sqrt{3} \rightarrow x = \frac{\pi}{3} + n\pi$ for all integers n OR

$\sec x = -2 \rightarrow \cos x = \frac{-1}{2} \rightarrow x = \frac{2\pi}{3} + 2n\pi$ OR $x = \frac{4\pi}{3} + 2n\pi$ for integers n.

4. $\csc^2 x - 2 = 0$

$\csc^2 x = 2$

$\csc x = \pm\sqrt{2}$

$\sin x = \pm\dfrac{1}{\sqrt{2}}$

$x = \dfrac{\pi}{4} + n\pi$ for all integers n.

5. $\cot x - \sqrt{3} = 0$

$\cot x = \sqrt{3}$

$\tan x \dfrac{1}{\sqrt{3}}$

$x = \dfrac{\pi}{6} + n\pi$ for all integers n.

6. $2\cos^2 x = \sin x + 1$

Switch everything to sine

$2(1 - \sin^2 x) = \sin x + 1$

$2 - 2\sin^2 x = \sin x + 1$ A quadratic!

$2\sin^2 x + \sin x - 1 = 0$ Factor

$(2\sin x - 1)(\sin x + 1) = 0$

$2\sin x - 1 = 0 \rightarrow \sin x = \dfrac{1}{2} \rightarrow x = \dfrac{\pi}{6} + 2n\pi, \dfrac{5\pi}{6} + 2n\pi$ for all integers n,   OR

$\sin x + 1 = 0 \rightarrow \sin x = -1 \rightarrow x = \dfrac{3\pi}{2} + 2n\pi$ for all integers n.

7. $3\sin 2x - 2\sin x = 0$

$3(2\sin x \cos x) - 2\sin x = 0$

$6\sin x \cos x - 2\sin x = 0$

$2\sin x(3\cos x - 1) = 0$

$2\sin x = 0 \rightarrow x = n\pi$ for all integers n OR

$3\cos x = 1 \rightarrow \cos x = \dfrac{1}{3}$ which a calculator is needed to solve.

8. $\tan^2 x - 2\sec x = 2$

$\sec^2 x - 1 - 2\sec x = 2$

$\sec^2 x - 2\sec x - 3 = 0$

$(\sec x - 3)(\sec x + 1) = 0$

$\sec x = 3 \rightarrow \cos x = \dfrac{1}{3}$ which we need a calculator to solve. OR

$\sec x = -1 \rightarrow \cos x = -1 \rightarrow x = \pi + 2n\pi$ for all integers n.

9. $\sec x \tan x - \cos x \cot x = \sin x$

$\dfrac{1}{\cos x} \dfrac{\sin x}{\cos x} - \cos x \dfrac{\cos x}{\sin x} = \sin x$

$\dfrac{\sin x}{\cos^2 x} - \dfrac{\cos^2 x}{\sin x} = \sin x$ Multiply by LCD $= \cos^2 x \sin x$

$(\cos^2 x \sin x)\left(\dfrac{\sin x}{\cos^2 x} - \dfrac{\cos^2 x}{\sin x} = \sin x \right)$

$\sin^2 x - \cos^4 x = \cos^2 x \sin^2 x$. Switch everything to cosine.

$1 - \cos^2 x - \cos^4 x = \cos^2 x(1 - \cos^2 x) = \cos^2 x - \cos^4 x$

$1 - 2\cos^2 x = 0$

$-2\cos^2 x = -1$

$\cos^2 x = \dfrac{1}{2}$

$\cos x = \pm \dfrac{1}{\sqrt{2}}$

$x = \dfrac{\pi}{4} + n\dfrac{\pi}{2}$ for all integers n.

10. $\sin^2 x + 3 = 7\cos^2 x$

$1 - \cos^2 x + 3 = 7\cos^2 x$

$4 = 8\cos^2 x$

$\cos^2 x = \dfrac{4}{8}$

$\cos x = \pm\sqrt{\dfrac{1}{2}} = \pm\dfrac{1}{\sqrt{2}}$

$x = \dfrac{\pi}{4} + \dfrac{n\pi}{2}$ for all integers n

N. Things that Look ALOT Like Trinomials

1. $(x+1)^2 - 5(x+1) - 6 = 0$

   Let $u = x + 1$, then our equation becomes

   $u^2 - 5u - 6 = 0$

   $(u-6)(u+1) = 0$ (Always double check your factoring. Did you factor to this: $(u-3)(u-2)$?

   $u - 6 = 0 \rightarrow u = 6$ OR

   $u + 1 = 0 \rightarrow u = -1$

   Since our goal is to solve for x, we need to return to u in terms of x

   $x + 1 = u = 6 \rightarrow x = 5$ OR

   $x + 1 = u = -1 \rightarrow x = -2$

2. $2e^{2x} + 3e^x - 2 = 0$

   Let $u = e^x$ and notice that $u^2 = (e^x)^2 = e^{2x}$. So we have,

   $2u^2 + 3u - 2 = 0$ which factors to

   $(2u - 1)(u + 2) = 0$ ZPP!

   $2u - 1 = 0 \rightarrow u = \frac{1}{2}$ OR

   $u + 2 = 0 \rightarrow u = -2$

   Going back to x, we get

   $e^x = \frac{1}{2}$ OR $e^x = -2$. And since $e^x > 0$ for all x, we have only

   $e^x = \frac{1}{2}$. Using the natural log to solve for x,

   $x = \ln \frac{1}{2}$

3. $\sin^2(x) + \sin(x) = 2$

   Let $u = \sin x$

   $u^2 + u = 2$

   $u^2 + u - 2 = 0$

   $(u + 2)(u - 1) = 0$

   $u + 2 = 0 \rightarrow u = -2$ OR $u - 1 = 0 \rightarrow u = 1$

   Now we go back to x

$\sin x = -2$ (never!) OR

$\sin x = 1 \to x = \dfrac{\pi}{2} + 2n\pi$ for all integers n.

4. $\sin^2 x = 2\sin x + 3$

   Let $u = \sin x$

   $u^2 = 2u + 3$

   $u^2 - 2u - 3 = 0$

   $(u - 3)(u + 1) = 0$

   $u - 3 = 0 \to u = 3$ OR

   $u + 1 = 0 \to u = -1$

   Back to x!

   $\sin x = 3$ (of course, never)

   $\sin x = -1 \to x = \dfrac{3\pi}{2} + 2n\pi$ for all integers n

5. $x^{1/3} + x^{1/6} - 2 = 0$

   Let $u = x^{1/6}$. We see that $u^2 = (x^{1/6})^2 = x^{2/6} = x^{1/3}$

   So this give us

   $u^2 + u - 2 = 0$

   $(u + 2)(u - 1) = 0$

   $u + 2 = 0 \to u = -2$ OR

   $u - 1 = 0 \to u = 1$

   $x^{1/6} = -2$ Can't happen! (why?)

   $x^{1/6} = 1 \to (x^{1/6})^6 = 1^6 \to x = 1$

6. $x^4 + x^2 = 2$

   Let $u = x^2$

   $u^2 + u - 2 = 0$

   $(u + 2)(u - 1) = 0$

   $u = -2 \to x^2 = -2$ NEVER!   OR

$$u = 1 \rightarrow x^2 = 1 \rightarrow x = \pm 1$$

O. Inequalities

1. $\dfrac{3}{x} < \dfrac{1}{2}$

$$\dfrac{3}{x} - \dfrac{1}{2} < 0$$

$$\dfrac{6}{2x} - \dfrac{x}{2x} < 0$$

$$\dfrac{6-x}{2x} < 0$$

Factors	$x < 0$	$0 < x < 6$	$x > 6$
$6 - x$	+	+	-
$2x$	-	+	+
$\dfrac{6-x}{2x}$	-	+	-

We see from the chart that our solution set is $(-\infty, 0) \cup (6, \infty)$

2. $2 \leq 2x - 1 < 3$

$$2 + 1 \leq 2x - 1 + 1 < 3 + 1$$

$$3 \leq 2x < 4$$

$$\dfrac{3}{2} \leq x < 2$$

3. $x^2 - 2x > 0$

$x(x - 2) > 0$ To the sign chart!

Factors	$x < 0$	$0 < x < 2$	$x > 2$
x	-	+	+
$x - 2$	-	-	+
$x(x-2)$	+	-	+

We see from the sign chart that our solution set is $(-\infty, 0) \cup (2, \infty)$

4. $(x + 2)(x - 3) < -4$

$$x^2 - x - 6 + 4 < 0$$

$$x^2 - x - 2 < 0$$

$$(x - 2)(x + 1) < 0$$

Factors	$x < -1$	$-1 < x < 2$	$x > 2$
$x - 2$	-	-	+
$x + 1$	-	+	+
$(x-2)(x+1)$	+	-	+

We see from the sign chart that our solution set is $(-1, 2)$

5. $\dfrac{2-x}{x+1} > 4$

$\dfrac{2-x}{x+1} - 4 > 0$

$\dfrac{2-x}{x+1} - \dfrac{4(x+1)}{x+1}$

$\dfrac{2-x-4x-4}{x+1} > 0$

$\dfrac{-5x-2}{x+1} > 0$ Divide by a negative and flip the inequality.

$\dfrac{5x+2}{x+1} < 0$ The sign chart!

Factors	$x < -1$	$-1 < x < -\dfrac{2}{5}$	$x > -\dfrac{2}{5}$
$5x + 2$	-	-	+
$x + 1$	-	+	+
$(5x+2)(x+1)$	+	-	+

We see from the sign chart that our solution set is $\left(-1, -\dfrac{2}{5}\right)$

6. $\dfrac{(2-x)(x+1)}{3-x} < 0$

Factors	$x < -1$	$-1 < x < 2$	$2 < x < 3$	$x > 3$
$2 - x$	+	+	-	-
$x + 1$	-	+	+	+
$3 - x$	+	+	+	-
$\dfrac{(2-x)(x+1)}{3-x}$	-	+	-	+

We see from the sign chart that our solution set is $(-\infty, 1) \cup (2, 3)$

7. $|2x - 1| < 3$

$-3 < 2x - 1 < 3$

$-2 < 2x < 4$

94

$-1 < x < 2$

8. $|3 - x| > 2$

$3 - x > 2 \quad 3 - x < -2$

$-x > -1 \to x < 1 \quad \text{OR} \quad 3 - x < -2 \to -x < -5 \to x > 5$

In interval notation, our solution set is $(-\infty, 1) \cup (5, \infty)$

9. $x^2 - 3x - 10 \geq 0$

$(x - 5)(x + 2) \geq 0$

Factors	$x < -2$	$-2 < x < 5$	$x > 5$
$x - 5$	-	-	+
$x + 2$	-	+	+
$(x - 5)(x + 2)$	+	-	+

Therefore, our solution set is $(-\infty, -2] \cup [5, \infty)$

10. $2x^2 + 3x - 3 < -1$

$2x^2 + 3x - 2 < 0$

$(2x - 1)(x + 2) < 0$

Factors	$x < -2$	$-2 < x < \dfrac{1}{2}$	$x > \dfrac{1}{2}$
$2x - 1$	-	-	+
$x + 2$	-	+	+
$(2x - 1)(x + 2)$	+	-	+

The solution set is $\left(-2, \dfrac{1}{2}\right)$

P. Finding Equations of Lines (this will happen every day in calculus) and a bonus.

    (a) Find the equation of the line passing through the point $(2, 3)$ with slope $m = -2$

        Using the point slope formula of the line $y - y_1 = m(x - x_1)$

        $y - 3 = -2(x - 2) \leftarrow$ this is an appropriate form unless otherwise stated.

        Slope-intercept form:

        $y - 3 = -2x + 4$

        $y = -2x + 7$

(b) Find the equation of the line passing through the points $(4, 5)$ and $(-2, 7)$

We need first find the slope

$$m = \frac{7-5}{-2-4} = \frac{2}{-6} = -\frac{1}{3}$$

Now use the point-slope formula of a line using either point.

$$y - 5 = -\frac{1}{3}(x - 4)$$

(c) Find the equation of the line passing through $(-2, 4)$ and parallel to the line $2x - 3y = 7$

First find the slope of the line given:

$$-3y = -2x + 7$$

$$y = \frac{-2}{-3}x + \frac{7}{-3} = \frac{2}{3}x - \frac{7}{3} \rightarrow m = \frac{2}{3}$$

Since parallel lines have the same slope, use the point-slope formula with this slope and our given point.

$$y - 4 = \frac{2}{3}(x - -2)$$

$$y - 4 = \frac{2}{3}(x + 2)$$

(d) Find the equation of the line passing through $(3, -1)$ and perpendicular to the line

$$3x + 7y = -1$$

First find the slope of the line given:

$$7y = -3x - 1$$

$$y = \frac{-3}{7}x - \frac{1}{7} \rightarrow m = -\frac{3}{7}$$

Since the slope of a line perpendicular to a line with slope m is $-\frac{1}{m}$, the slope we want to use in the point slope formula is $m_2 = \frac{7}{3}$

$$y - -1 = \frac{7}{3}(x - 3)$$

$$y + 1 = \frac{7}{3}(x - 3)$$

Keep going...

Bonus

Find the equation of the circle with center $(-2, 3)$ and radius $r = 2$

$$(x - h)^2 + (y - k)^2 = r^2$$
$$(x - -2)^2 + (y - 3)^2 = 2^2$$
$$(x + 2)^2 + (y - 3)^2 = 4$$

www.ingramcontent.com/pod-product-compliance
Lightning Source LLC
Chambersburg PA
CBHW080643180526
45168CB00008B/3283